# Process Development Symposium 2021

Online
8-10 June 2021

ISBN: 978-1-7138-3705-3

**Printed from e-media with permission by:**

Curran Associates, Inc.
57 Morehouse Lane
Red Hook, NY 12571

**Some format issues inherent in the e-media version may also appear in this print version.**



Printed with permission by Curran Associates, Inc. (2021)

For permission requests, please contact AIChE
at the address below.

AIChE
120 Wall Street, FL 23
New York, NY 10005-4020

Phone: (800) 242-4363
Fax: (203) 775-5177

www.aiche.org

**Additional copies of this publication are available from:**

Curran Associates, Inc.
57 Morehouse Lane
Red Hook, NY 12571 USA
Phone: 845-758-0400
Fax: 845-758-2633
Email: curran@proceedings.com
Web: www.proceedings.com

# TABLE OF CONTENTS

## DIGITAL TWINS (OPERATIONS, PROCESS OPTIMIZATION & SAFETY)



## KEYNOTE ADDRESS



## DRIVING INNOVATION IN PROCESS DEVELOPMENT



## DATA ANALYTICS IN PROCESS DEVELOPMENT

**INTERACTIVE PANEL SESSION**



**ENGINEERING EDUCATION**



**ARTIFICIAL INTELLIGENCE**



**EFFICIENT PROCESS DEVELOPMENT MANAGEMENT**

## DEVELOPMENT IN THE AGE OF TECH TRANSFERS AND OUTSOURCING



## 3D PRINTING



## SAFETY IN PROCESS DEVELOPMENT



## SCALE-UP & SCALE-DOWN - RISK MANAGEMENT FOR ACCELERATED PROCESS DEVELOPMENT

## BATCH VS CONTINUOUS

# Numerical Twins of Phyisical Systems

**Conference:** Process Development Symposium

**Year:** 2021

**Proceeding:** 2021 Process Development Symposium

**Group:** General Submissions

**Session:** Digital Twins (Operations, Process Optimization & Safety)

**Time:** Tuesday, June 8, 2021 - 8:15am-8:30am

## Authors:

**Thomas, J. A. - Presenter**, M-Star Simulations

The enrichment and sintering plant of the Industrial Complex of Moanda (CIM) enables the valorization of Mn-rich fines generated during the production of MMA and MMD ores.

Thanks to the analysis of historical data of the sintering plant, a data-driven Digital Twin of the process was developed. The goal behind the development of this Digital Twin is to precisely mimic the behavior of the industrial process and enable the identification of optimal setpoints in the sintering process.

Certain key parameters of the process, such as the permeability of the load and its temperature during sintering, are nevertheless difficult to measure online and therefore missing in the digital twin. To gain information on these parameters, a First-Principle model of the sintering strand was developed. Such a physics-driven model enriches the Digital Twin to deduce the behavior of the process beyond its conventional operating ranges. Based on kinetics of coke combustion, thermodynamics of the sintering reaction, and aeraulics of the gas flow within the load, the advance of the flame front and the temperature profile in the load were simulated. When run in an inverse manner coupled to an optimization algorithm, the model enabled the calculation of the load's permeability. Thereafter, the permeability values were used to enrich the Digital Twin.

The Digital Twin and the physical model have thus been interconnected. This provides a powerful tool for optimizing CIM performance. In the future, this digital tool developed at Eramet Ideas will be deployed in the plant. On-site missions will allow us to receive user feedback and work directly on improving the user interface.

# Digital Twin Applied on Crystallisation Process

**Conference:** Process Development Symposium

**Year:** 2021

**Proceeding:** 2021 Process Development Symposium

**Group:** General Submissions

**Session:** Digital Twins (Operations, Process Optimization & Safety)

**Time:** Tuesday, June 8, 2021 - 7:50am-8:15am

## Authors:

**Henriques, J. - Presenter**, Inevo Technologies

The enrichment and sintering plant of the Industrial Complex of Moanda (CIM) enables the valorization of Mn-rich fines generated during the production of MMA and MMD ores.

Thanks to the analysis of historical data of the sintering plant, a data-driven Digital Twin of the process was developed. The goal behind the development of this Digital Twin is to precisely mimic the behavior of the industrial process and enable the identification of optimal setpoints in the sintering process.

Certain key parameters of the process, such as the permeability of the load and its temperature during sintering, are nevertheless difficult to measure online and therefore missing in the digital twin. To gain information on these parameters, a First-Principle model of the sintering strand was developed. Such a physics-driven model enriches the Digital Twin to deduce the behavior of the process beyond its conventional operating ranges. Based on kinetics of coke combustion, thermodynamics of the sintering reaction, and aeraulics of the gas flow within the load, the advance of the flame front and the temperature profile in the load were simulated. When run in an inverse manner coupled to an optimization algorithm, the model enabled the calculation of the load's permeability. Thereafter, the permeability values were used to enrich the Digital Twin.

The Digital Twin and the physical model have thus been interconnected. This provides a powerful tool for optimizing CIM performance. In the future, this digital tool developed at Eramet Ideas will be deployed in the plant. On-site missions will allow us to receive user feedback and work directly on improving the user interface.

# Anomaly Detection Approaches Applied to Process

**Conference:** Process Development Symposium

**Year:** 2021

**Proceeding:** 2021 Process Development Symposium

**Group:** General Submissions

**Session:** Digital Twins (Operations, Process Optimization & Safety)

**Time:** Tuesday, June 8, 2021 - 8:30am-8:55am

## Authors:

**Firdova, K. - Presenter**, Optimistik

The enrichment and sintering plant of the Industrial Complex of Moanda (CIM) enables the valorization of Mn-rich fines generated during the production of MMA and MMD ores.

Thanks to the analysis of historical data of the sintering plant, a data-driven Digital Twin of the process was developed. The goal behind the development of this Digital Twin is to precisely mimic the behavior of the industrial process and enable the identification of optimal setpoints in the sintering process.

Certain key parameters of the process, such as the permeability of the load and its temperature during sintering, are nevertheless difficult to measure online and therefore missing in the digital twin. To gain information on these parameters, a First-Principle model of the sintering strand was developed. Such a physics-driven model enriches the Digital Twin to deduce the behavior of the process beyond its conventional operating ranges. Based on kinetics of coke combustion, thermodynamics of the sintering reaction, and aeraulics of the gas flow within the load, the advance of the flame front and the temperature profile in the load were simulated. When run in an inverse manner coupled to an optimization algorithm, the model enabled the calculation of the load's permeability. Thereafter, the permeability values were used to enrich the Digital Twin.

The Digital Twin and the physical model have thus been interconnected. This provides a powerful tool for optimizing CIM performance. In the future, this digital tool developed at Eramet Ideas will be deployed in the plant. On-site missions will allow us to receive user feedback and work directly on improving the user interface.

# Accelerated Chemical Reaction Optimization with Active Learning to Build your Reaction Digital Twin

**Conference:** Process Development Symposium

**Year:** 2021

**Proceeding:** 2021 Process Development Symposium

**Group:** General Submissions

**Session:** Digital Twins (Operations, Process Optimization & Safety)

**Time:** Tuesday, June 8, 2021 - 7:00am-7:25am

## Authors:

**Galeandro-Diamant, T. - Presenter**, ChemIntelligence

The optimization of chemical reactions is very often performed using a trial-and-error methodology and modifying one experimental variable at a time. In this presentation, we will show how the active learning methodology is able to reduce the number of experiments required to optimize chemical reactions by building a "digital twin" of the chemical reaction system.

We will present one of the most well-known active learning algorithms: Bayesian optimization. This algorithms combines two strategies (exploration and exploitation) to build a surrogate machine learning model of the chemical reaction experiment-by-experiment. This surrogate model, the "digital twin" of our chemical reaction system, is combined with an acquisition function in order to iteratively and quickly find an optimal combination of reaction parameters.

We will discuss the advantages and challenges of optimizing chemical reactions using Bayesian optimization algorithms and compare this machine learning-based approach with the traditional Design of Experiments (DoE) methodology.

# A Hybrid Digital Twin of the Sintering Process of Manganese Ore

**Conference:** Process Development Symposium

**Year:** 2021

**Proceeding:** 2021 Process Development Symposium

**Group:** General Submissions

**Session:** Digital Twins (Operations, Process Optimization & Safety)

**Time:** Tuesday, June 8, 2021 - 7:25am-7:50am

## Authors:

**Han, H. - Presenter**, Eramet Ideas
Vansteene, A., Eramet Ideas

The enrichment and sintering plant of the Industrial Complex of Moanda (CIM) enables the valorization of Mn-rich fines generated during the production of MMA and MMD ores.

Thanks to the analysis of historical data of the sintering plant, a data-driven Digital Twin of the process was developed. The goal behind the development of this Digital Twin is to precisely mimic the behavior of the industrial process and enable the identification of optimal setpoints in the sintering process.

Certain key parameters of the process, such as the permeability of the load and its temperature during sintering, are nevertheless difficult to measure online and therefore missing in the digital twin. To gain information on these parameters, a First-Principle model of the sintering strand was developed. Such a physics-driven model enriches the Digital Twin to deduce the behavior of the process beyond its conventional operating ranges. Based on kinetics of coke combustion, thermodynamics of the sintering reaction, and aeraulics of the gas flow within the load, the advance of the flame front and the temperature profile in the load were simulated. When run in an inverse manner coupled to an optimization algorithm, the model enabled the calculation of the load's permeability. Thereafter, the permeability values were used to enrich the Digital Twin.

The Digital Twin and the physical model have thus been interconnected. This provides a powerful tool for optimizing CIM performance. In the future, this digital tool developed at Eramet Ideas will be deployed in the plant. On-site missions will allow us to receive user feedback and work directly on improving the user interface.

# Key Insights for success in Rapid Scale-Up of Chemical Processes

**Conference:** Process Development Symposium

**Year:** 2021

**Proceeding:** 2021 Process Development Symposium

**Group:** General Submissions

**Session:** 2021 Process Development Symposium Keynote Address: Key Insights for success in Rapid Scale-Up of Chemical Processes

**Time:** Tuesday, June 8, 2021 - 9:00am-10:00am

## Authors:

**Dems, B. - Presenter**, Global Leader, Leveraged R&D and Emerging Technologies, DuPont

The desire and need for a team to rapidly develop and scale-up a new and novel chemical process can be an invigorating and rewarding experience if done well and comprehensively or be a source of grief and disappointment when key elements of the technology development are missed or fail to deliver at desired levels of product performance & consistency or production rate. This talk will summarize some of the most salient features of successful rapid process developments & scale-ups and major lessons-learned over three decades of personal experience. An over-arching theme of identifying early-on the major Critical To Quality (CTQ) specifications for end-product and product throughput rates is critical to success. The need to identify the top few technology-related critical issues and assembling cross-functional problem-solving teams to develop options and solutions using a variety of techniques and resources will be discussed.

# Statistics & Machine Learning in Chemical Processes

**Conference:** Process Development Symposium

**Year:** 2021

**Proceeding:** 2021 Process Development Symposium

**Group:** General Submissions

**Session:** Driving Innovation in Process Development

**Time:** Tuesday, June 8, 2021 - 11:00am-11:30am

## Authors:

**Castillo, F. - Presenter**, SABIC
Lapanowski, A., SABIC

The desire and need for a team to rapidly develop and scale-up a new and novel chemical process can be an invigorating and rewarding experience if done well and comprehensively or be a source of grief and disappointment when key elements of the technology development are missed or fail to deliver at desired levels of product performance & consistency or production rate. This talk will summarize some of the most salient features of successful rapid process developments & scale-ups and major lessons-learned over three decades of personal experience. An over-arching theme of identifying early-on the major Critical To Quality (CTQ) specifications for end-product and product throughput rates is critical to success. The need to identify the top few technology-related critical issues and assembling cross-functional problem-solving teams to develop options and solutions using a variety of techniques and resources will be discussed.

# RAPID Manufacturing Institute™ - Transforming the Process Industries through Consortium-Based Innovation

**Conference:** Process Development Symposium

**Year:** 2021

**Proceeding:** 2021 Process Development Symposium

**Group:** General Submissions

**Session:** Driving Innovation in Process Development

**Time:** Tuesday, June 8, 2021 - 12:15pm-12:45pm

## Authors:

**Grieco, W. - Presenter**, CEO, RAPID Manufacturing Institute , AIChE
Palou-Rivera, I., AIChE

Digital process design involves far more than just the process simulation typically used to design and optimize processes. It comprises an integrated set of techniques and methodologies that deploy validated digital twins of processes to explore the decision space rapidly and effectively, optimize the process and product designs and quantify and manage technology risk.

This presentation describes how Siemens Process Systems Engineering digital design tools and methodologies are used across a spectrum of process industries and highlights the key technology that enables this. Benefits of this approach include accelerated innovation and faster time-to-market with better products and process performance, which is all achieved with reduced risk and dependence on costly experimentation.

# Process Development and Scale-Up of a Photochemical Reaction for Drug Substance Intermediate

**Conference:** Process Development Symposium

**Year:** 2021

**Proceeding:** 2021 Process Development Symposium

**Group:** General Submissions

**Session:** Driving Innovation in Process Development

**Time:** Tuesday, June 8, 2021 - 11:30am-12:00pm

## Authors:

**McMullen, J. P. - Presenter**, Merck & Co. Inc.

Recent advances in light-emitting diode (LED) technology and development of visible-light photoredox catalysis have led to a resurgence of interest in photochemical transformations for drug substance development. Photochemical approaches can streamline the manufacturing of active pharmaceutical ingredient (API) by providing reaction selectivity that cannot be achieved with traditional synthetic methods. Despite the intrinsic synthetic value of these approaches, applications in the pharmaceutical industry have largely been focused in the early stages of drug development, with very few examples of photochemistry implemented beyond lab scale.

This presentation will discuss various strategies that have been used to develop and scale up a photochemical bromination. Small, benchtop experiments were used to identify the reaction mechanism and understand reaction sensitivities. By establishing photon equivalence as the scaling factor, the bromination was scaled up to larger batch reactors and tech transferred to continuous flow reactors. The process was scaled up to the kilo-scale with two different flow reactor configurations to understand the impact of reactor design on photochemical reaction performance before selecting a plug flow reactor for manufacturing. To further understand the bromination reaction in a production environment, the process was explored at various residence times, irradiation levels, and temperatures in the pilot plant. This data, along with kilo-scale process characterization investigations, enabled successful tech transfer of the process to manufacturing to produce metric tons of drug substance intermediate.

# Digital Process Design - using validated high-fidelity Digital Process Twins to accelerate innovation and manage risk

**Conference:** Process Development Symposium

**Year:** 2021

**Proceeding:** 2021 Process Development Symposium

**Group:** General Submissions

**Session:** Driving Innovation in Process Development

**Time:** Tuesday, June 8, 2021 - 12:00pm-12:15pm

## Authors:

**Leyland, S. - Presenter**, Process Systems Enterprise

Digital process design involves far more than just the process simulation typically used to design and optimize processes. It comprises an integrated set of techniques and methodologies that deploy validated digital twins of processes to explore the decision space rapidly and effectively, optimize the process and product designs and quantify and manage technology risk.

This presentation describes how Siemens Process Systems Engineering digital design tools and methodologies are used across a spectrum of process industries and highlights the key technology that enables this. Benefits of this approach include accelerated innovation and faster time-to-market with better products and process performance, which is all achieved with reduced risk and dependence on costly experimentation.

# Use of Advanced Multivariate Analysis to Solve Plant Problems

**Conference:** Process Development Symposium

**Year:** 2021

**Proceeding:** 2021 Process Development Symposium

**Group:** General Submissions

**Session:** Data Analytics in Process Development

**Time:** Tuesday, June 8, 2021 - 1:30pm-2:00pm

## Authors:

**Somasi, S. - Presenter**, Dow Chemical Company

Cross-functional collaboration of statisticians and chemical engineers can be a winning strategy to solve chemical plant problems. In modern chemical engineering plants, we collect massive amounts of data. This plant data is highly correlated due to the underlying mass and energy balance. Multivariate analyses, often referred to as chemometrics, have been used to analyze highly correlated data to solve plant problems. This data analysis combined with the right subject matter expertise can be a powerful strategy for troubleshooting. In this presentation, we will give an overview of multivariate analysis using a commercial tool (SIMCA). We will provide examples using actual plant troubleshooting problems. The examples will highlight how using these advanced data analytics methods, we were able to compare the stable plant operating conditions with the problematic ones and quickly identify the main differences. This led to identifying the exact root cause and quick resolution of the issue.

# Facilitate Process Optimization using Machine Learning Methodologies

**Conference:** Process Development Symposium

**Year:** 2021

**Proceeding:** 2021 Process Development Symposium

**Group:** General Submissions

**Session:** Data Analytics in Process Development

**Time:** Tuesday, June 8, 2021 - 2:00pm-2:30pm

## Authors:

**Wang, Z. - Presenter**, Dow Inc.

Process optimization is one of the key components for process development. Due to the complex nature of the industrial processes, especially (semi-) batches, process optimization usually relies on knowledge-driven models, which are developed based on first principles. In the last 50 years, the application of knowledge-driven models has increased considerably, especially for continuous chemical, petroleum, and petrochemical processes1. However, developing an accurate, knowledge-driven model is not always achievable. Some challenges that engineers/scientists may frequently encounter are: 1) lack of knowledge about process inner workings; 2) high cost to collect informative data to improve the model; 3) difficulty to justify the return from investment for model development; and 4) long development time of fundamental model to achieve the process optimization task.

As an alternative, machine learning methodologies (also known as data-driven methodologies) that optimize the process-based input-output relationships could be applied to optimize processes in a timely and economical manner. It has been proved that with a few well-designed experiments, similar optimal process performance could be achieved using a data-driven approach compared to the optimum obtained using knowledge-driven models2. The limitation of the data-driven methodologies is that their interpretability is lower compared to knowledge-driven approaches. So, leveraging the advantages of both data-driven and knowledge-driven approaches to facilitate the process optimization tasks is of high priority.

In this paper, we propose a hybrid methodology that combines an existing yet not very accurate knowledge-driven model with historical data collected from regular manufacturing activities to optimize the performance of a semi-batch process without running new experiments. Specifically, a Long-Short-Term-Memory (LSTM) network3-4 is developed using both simulated data from the knowledge-driven model and the historical data. The LSTM model learns the inner workings of the process described by both datasets. Then the LSTM network serves as an interactive environment to train the reinforcement learning (RL) agent5 6 to learn optimized fed-batch profiles. The RL agent suggests new optimal conditions with higher reactor temperature and higher reactant feed than the current ones. A 3% improvement in product values is suggested after implementing the optimal conditions in the knowledge-driven model. In addition, the estimated improvement also helps to justify the investment to further develop a more accurate and complete first principle model that could lead to higher process improvement.

Reference

1. Bonvin, D.; Georgakis, C.; Pantelides, C.; Barolo, M.; Grover, M.; Rodrigues, D.; Schneider, R.; Dochain, D., Linking models and experiments. Industrial & Engineering Chemistry Research 2016, 55 (25), 6891-6903.

2. Georgakis, C.; Chin, S.-T.; Wang, Z.; Hayot, P.; Chiang, L. H.; Wassick, J. M.; Castillo, I., Data-Driven Optimization of an Industrial Batch Polymerization Process Using the Design of Dynamic Experiments Methodology. Industrial & Engineering Chemistry Research 2020.

3. Hochreiter, S.; Schmidhuber, J., Long short-term memory. Neural computation 1997, 9 (8), 1735-1780.

4. Ma, Y.; Wang, Z.; Castillo, I.; Rendall, R.; Bindlish, R.; Aschraft, B.; Bentley, D.; Benton, M.; Romagnoli, J. A.; Chiang, L. In Reinforcement Learning-Based Fed-Batch Optimization with Reaction Surrogate Model, American Control Conference (ACC), New Orleans, Louisiana, USA, New Orleans, Louisiana, USA, 2021.

5. Schulman, J.; Wolski, F.; Dhariwal, P.; Radford, A.; Klimov, O., Proximal policy optimization algorithms. arXiv preprint arXiv:1707.06347 2017.

6. Rendall , R.; Ma, Y.; Castillo, I.; Wang, Z.; Peng, Y.; Chiang, L. In Applying Reinforcement Learning for Batch Trajectory Optimization in a Chemical Industrial Applicatio, AIChE Spring Meeting, Virtual, Virtual, 2021.

# Accelerating Multiscale Process Design with Bayesian Optimization: Progress, Challenges, and Opportunities

**Conference:** Process Development Symposium

**Year:** 2021

**Proceeding:** 2021 Process Development Symposium

**Group:** General Submissions

**Session:** Data Analytics in Process Development

**Time:** Tuesday, June 8, 2021 - 1:00pm-1:30pm

## Authors:

**Paulson, J. - Presenter**, The Ohio State University

Design problems, which are pervasive in science, engineering, and manufacturing endeavors, can be generally formulated as mathematical optimization problems. In ideal situations, one is able to develop a physics-based model of the system of interest whose structure can be exploited by state-of-the-art solvers, especially in cases that the derivatives can be exactly computed. However, the development of multiscale process models for which derivative information is readily available remains a significant challenge in many real-world applications. A particularly challenging class of problems is when at least one component of the model is very expensive or time-consuming to evaluate such as high-fidelity computer simulations (e.g., thermodynamic calculations, molecular dynamic simulations, or solutions to partial differential equations) and laboratory experiments (e.g., measurement of critical material, chemical, or biological properties). Although "black-box" optimization methods can be applied in these situations, many of the available algorithms require extensive sampling and thus are not tractable for expensive-to-evaluate objectives and constraints.

Recently, Bayesian optimization (BO) has emerged as a powerful tool for optimizing expensive black-box functions due to its successes in material/drug design and hyperparameter optimization in machine learning algorithms. In this talk, we provide an overview of BO and discuss its main advantages and disadvantages in the context of process systems applications. We also discuss two new advances in BO that can deliver considerable gains in performance by effectively "peeking inside the box" (i.e., selectively exploiting problem structure whenever possible), which we refer to as "grey-box" BO methods. In particular, we show how BO can be modified to handle composite functions and functions for which multiple lower-fidelity (cheap-to-evaluate) approximations are available. We then describe applications of these methods to (i) integrated design and control of flexible building heating and cooling systems with hourly variation in weather conditions over year-long simulations and (ii) calibrating genome-scale bioreactor models to experimental data.

# Accelerated Process Development & Digitization for Commercial Success

**Conference:** Process Development Symposium

**Year:** 2021

**Proceeding:** 2021 Process Development Symposium

**Group:** General Submissions

**Session:** Interactive Panel Session

**Time:** Tuesday, June 8, 2021 - 2:30pm-3:30pm

## Authors:

**Sarafinas, A. - Presenter**, Sarafinas Process & Mixing Consulting LLC

Since the beginning of the 21st century, the chemical industry has been making many novel and ingenious innovations to design cost-advantaged, inherently safe, and sustainable processes based on renewably- sourced feedstocks for the production of fuels, chemicals, and advanced materials. However, there is still a persistent public perception that the chemical industry is not doing enough to develop safe and sustainable chemistry. As a matter of fact, a large number of scientists, environmental scholars, business leaders, and governmental policymakers believe that chemical industry must do more to develop sustainable processes to save our planet from the dire consequences of climate change.

During the last three decades, DuPont has been in the forefront of developing sustainable products and processes. One such endeavor was to develop a cost-competitive process for the synthesis of 1,6-hexanediol from renewable feedstocks. Several potential feedstocks have been investigated with the main intermediates being 1,2,6-hexanetriol (1,2,6-HT) and tetrahydropyran-2-methanol (THP2M) as shown in Figure 1.

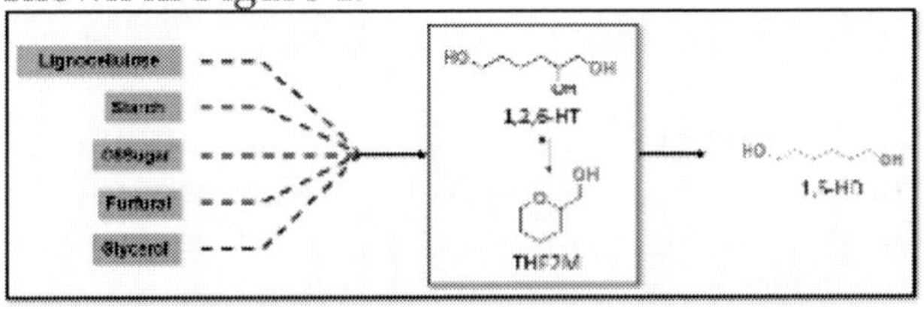


Figure 1: Reactions highlighting the main intermediates for converting biomass-derived feedstocks to renewable 1,2-hexanediol

Multimetallic catalysts have been shown to be valuable for a variety of industrially-practiced reactions, including biomass valorization. We have explored the activity, selectivity, and life of supported metal-metal oxide catalyst combinations using high throughput screening and continuous trickle bed reactors for the hydrodeoxygenation of biomass-derived feedstocks. [1-5]

The development work started with the synthesis and screening of plethora of supported metal and metal-oxide catalyst systems (Figure 2). The initial proof of concept (PoC) was performed in small high- throughput batch reactors (milligram quantity of catalysts in powder form). After completing PoC, the process was transferred from batch reactors to continuous reactors, which comprised of high-throughput trickle-bed test units (gram catalyst scale, in powder form, with 16 reactor tubes in

parallel) to closely match the operating conditions of a commercial continuous hydrodeoxygenation process. The first two steps comprised of screening several hundred catalysts under a broad range of operating conditions to determine the activity, selectivity, and stability of the catalysts. Performance targets were set based on the initial results and techno-economic evaluation. In the next step, several lead catalyst candidates were identified and subsequently scaled-up to shapes (extrudates and pellets) and sizes (e.g., 1/16â€ and 1/8â€ )
similar to that of standard commercial catalysts. During this evaluation process, a few grams of the selected catalysts were further optimized by extending the range of operating conditions. In the final stage of the development work, the amount of catalyst was increased into hundred-gram scale using a semiworks-scale test rig. The latter had a lower degree of parallelization but a high degree of flexibility for each catalyst system. The catalyst systems were operated independently in their preferred reaction conditions and over a longer period with industrially relevant feedstock to assess long-term catalyst stability. Larger quantities of the effluent streams were collected for more detailed product analytics and separation and purification workstreams. The results obtained from the reactor scale-up study were used to further refine the techno-economic analysis and were used as an input for designing the pilot plant.

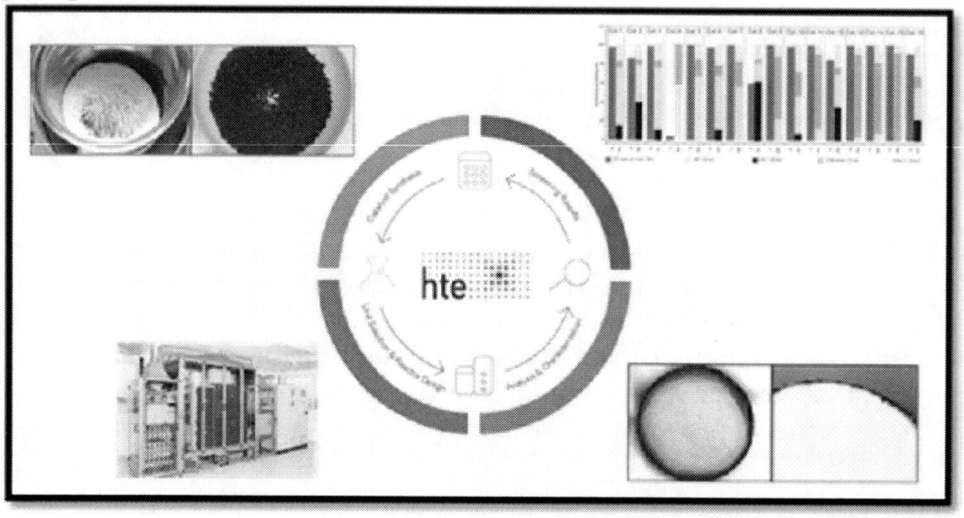


Figure 2: Development cycle for producing 1,6-HD from C6 polyols. The catalyst scale-up in this example involved shell impregnation of commercial supports. The product analysis involved detailed GC-MS and GC-FID/TCD analyses of the gas- and liquid-phase products. The testing included batch and continuous high throughput screening and flexible semiworks-scale reactor system.

The performance of the catalysts, with special reference to Pt-WOx/TiO2, the lead catalyst candidate, for the hydrodeoxygenation of 1,2-HT and THP2M and the significance of the batch and continuous reactor systems for the synthesis of 1,6-HD will be discussed in the presentation.

# References

1. Allgeier, A. M., Carlson; T. R., Corbin, D. R., De Silva, W. I. N., Menning, C. A., Ritter, J. C., and Sengupta, S. K., Production of alpha, omega-diols US 9,670,118 (2017)
2. Allgeier, A. M., Carlson; T. R., Corbin, D. R., De Silva, W. I. N., Korovessi, E., Menning, C. A., Ritter, J. C., Rosenfeld; H. D. and Sengupta, S. K., Production of alpha, omega-diols US 9,181,157 (2015)
3. Stauffer, C. S., Sengupta, S. K., Ritter, J. C., Menning, C. A., Korovessi, E., De Silva, W. I. N., Allgeier,
A. M., Process for Preparing 1,6-Hexanediol, US 8,865,940 (2014)
4. Stauffer, C. S., Sengupta, S. K., Ritter, J. C., Menning, C. A., Korovessi, E., De Silva, W. I. N., Allgeier, A. M., Process for Preparing 1,6-Hexanediol, US 8,865,940 (2014)

5. Stephens, K. J., Allgeier, A. M., Bell, A. L., Carlson, T. R., Chen, Y., Douglas, J. T., Howe, L. A., Menning, C. A., Neuenswander, S. A., Sengupta, S. K., Thapa, and Ritter, J. C. P. S. A Mechanistic Study of Polyol Hydrodeoxygenation over a Bifunctional Pt-WOX/TiO2 Catalyst, ACS Catal. 10, 12996â^'13007 (2020)

# The cloud's silver lining: Active learning in pandemic times

**Conference:** Process Development Symposium

**Year:** 2021

**Proceeding:** 2021 Process Development Symposium

**Group:** General Submissions

**Session:** Engineering Education

**Time:** Wednesday, June 9, 2021 - 7:30am-8:00am

## Authors:

**Liauw, M. A. - Presenter**, ITMC, RWTH Aachen University

Industry 4.0 is now a well-established term that chemical and process industries strive towards. Educating future (bio)chemical engineers that have the necessary knowledge and skills in this area is, however, still seen as a challenge, particularly in the current pandemic-affected era in higher education. This contribution will highlight how the teaching of these important topics will need to be modified in the new context of limited face to face delivery with social distancing. It will highlight the main challenges and compare the mode of delivery pre and post-Covid-19 and touch on the important issues of ensuring high quality learning experience for the students. Case studies exemplifying the use of immersive technologies, such as augmented and virtual reality, will raise some future educational prospects.

# Digitalisation and advanced modelling skills in chemical engineering education

**Conference:** Process Development Symposium

**Year:** 2021

**Proceeding:** 2021 Process Development Symposium

**Group:** General Submissions

**Session:** Engineering Education

**Time:** Wednesday, June 9, 2021 - 7:00am-7:30am

## Authors:

**Glassey, J. - Presenter**, Newcastle University

Industry 4.0 is now a well-established term that chemical and process industries strive towards. Educating future (bio)chemical engineers that have the necessary knowledge and skills in this area is, however, still seen as a challenge, particularly in the current pandemic-affected era in higher education. This contribution will highlight how the teaching of these important topics will need to be modified in the new context of limited face to face delivery with social distancing. It will highlight the main challenges and compare the mode of delivery pre and post-Covid-19 and touch on the important issues of ensuring high quality learning experience for the students. Case studies exemplifying the use of immersive technologies, such as augmented and virtual reality, will raise some future educational prospects.

# Chemical Engineering Education in the Digital age: Opportunities and Challenges

**Conference:** Process Development Symposium

**Year:** 2021

**Proceeding:** 2021 Process Development Symposium

**Group:** General Submissions

**Session:** Engineering Education

**Time:** Wednesday, June 9, 2021 - 8:00am-8:30am

## Authors:

**Ventura-Medina, E. - Presenter**, University of Strathclyde

In the last few years there has been a tremendous change in the education landscape mainly due to the rapid technological developments around us. Over time universities have been adopting the use of new technologies in particular to complement on-campus activities. The use of Virtual Learning Environments has become a necessity for teaching and learning delivery. Chemical Engineering programmes have also engaged significantly in this 'digital transformation' and never more so than over the past year. Even if there had been reluctance to change, the Covid19 pandemic has forced the chemical engineering education establishment to adapt and evolve. Nevertheless, how much change has actually occurred?

Education is not the only sector that has been affected by digitalisation. It is relevant therefore to distinguish between the digital transformation in education and that which has occurred in engineering practice. For a student to operate in the education environment it is important to have enough 'digital literacy'. However, for a graduate to operate in the work place it is important to have 'digital proficiency'. Given that the needs of both sectors are different, a pertinent question is how changes in education delivery have impacted on the development of 'digital proficiency' in chemical engineering graduates.

In this presentation I will discuss some of the opportunities that recent changes education environments have opened up, what are the affordances that digital technologies offer for education of chemical engineering graduates and what are the gaps that still need to be breached between 'training/education' and 'practice'. I will also discuss the challenges that have arisen and the difficulties that lie ahead.

# Leveraging AI for the Future of Sustainability Innovation

**Conference:** Process Development Symposium

**Year:** 2021

**Proceeding:** 2021 Process Development Symposium

**Group:** General Submissions

**Session:** Artificial Intelligence

**Time:** Wednesday, June 9, 2021 - 10:00am-10:30am

## Authors:

**van der Heijden, T. - Presenter**, Merantix

The availability of sufficient plant data, both in quality and quantity, is a barrier to widespread usage of Machine Learning (ML) models in the process industry. Flowsheet simulators can solve this problem effectively by providing reliable data that considers the underlying physical models. They can also act as evaluation and validation tools for ML models. The resulting hybrid models unleash a true potential for ML applications by combining data-driven predictions with mass and energy balances. The ML model then becomes an attractive alternative to rigorous simulation, which can solve more complex process engineering problems in lower calculation times.

In this presentation, we introduce our 5-step workflow for building and evaluating ML models based on rigorous simulation models. We build an ML model for a pre-reformer reactor and evaluate it by implementing the model as a unit operation.

# Divide and conquer: Benefits from modular surrogate modelling of flowsheets

**Conference:** Process Development Symposium

**Year:** 2021

**Proceeding:** 2021 Process Development Symposium

**Group:** General Submissions

**Session:** Artificial Intelligence

**Time:** Wednesday, June 9, 2021 - 9:30am-10:00am

## Authors:

**Bortz, M. - Presenter,** Fraunhofer Institute for Industrial Mathematics (ITWM)

The availability of sufficient plant data, both in quality and quantity, is a barrier to widespread usage of Machine Learning (ML) models in the process industry. Flowsheet simulators can solve this problem effectively by providing reliable data that considers the underlying physical models. They can also act as evaluation and validation tools for ML models. The resulting hybrid models unleash a true potential for ML applications by combining data-driven predictions with mass and energy balances. The ML model then becomes an attractive alternative to rigorous simulation, which can solve more complex process engineering problems in lower calculation times.

In this presentation, we introduce our 5-step workflow for building and evaluating ML models based on rigorous simulation models. We build an ML model for a pre-reformer reactor and evaluate it by implementing the model as a unit operation.

# AI-improved soft sensor for the separation efficiency of a packed column

**Conference:** Process Development Symposium

**Year:** 2021

**Proceeding:** 2021 Process Development Symposium

**Group:** General Submissions

**Session:** Artificial Intelligence

**Time:** Wednesday, June 9, 2021 - 9:00am-9:30am

## Authors:

**Schöneberger, J. - Presenter**, Chemstations Europe GmbH

The availability of sufficient plant data, both in quality and quantity, is a barrier to widespread usage of Machine Learning (ML) models in the process industry. Flowsheet simulators can solve this problem effectively by providing reliable data that considers the underlying physical models. They can also act as evaluation and validation tools for ML models. The resulting hybrid models unleash a true potential for ML applications by combining data-driven predictions with mass and energy balances. The ML model then becomes an attractive alternative to rigorous simulation, which can solve more complex process engineering problems in lower calculation times.

In this presentation, we introduce our 5-step workflow for building and evaluating ML models based on rigorous simulation models. We build an ML model for a pre-reformer reactor and evaluate it by implementing the model as a unit operation.

# Translating Innovation to Solve Real-World Problems : Commercialization Process

**Conference:** Process Development Symposium

**Year:** 2021

**Proceeding:** 2021 Process Development Symposium

**Group:** General Submissions

**Session:** Efficient Process Development Management

**Time:** Wednesday, June 9, 2021 - 11:00am-11:30am

## Authors:

**Krishnan, B. - Presenter**, The Dow Chemical Company

Innovation and Technology drives the commercialization process that results in product leadership and market competitiveness. At the same time, product commercialization is essential for the growth and sustainability of any industry. As a result, industries are adapting to a fast paced approach to product development followed by an effective and streamlined commercialization process. These processes involve developments and activities connecting different business functions. This talk provides insight into the processes involved, the challenges, the partnerships, and other related activities pertaining to the progression from Idea to a product to commercialization.

# The Role of Technology Risk in Guiding Efficient Process Development

**Conference:** Process Development Symposium

**Year:** 2021

**Proceeding:** 2021 Process Development Symposium

**Group:** General Submissions

**Session:** Efficient Process Development Management

**Time:** Wednesday, June 9, 2021 - 11:30am-12:00pm

## Authors:

**Trippeer, M. - Presenter**, Corteva

Innovation and Technology drives the commercialization process that results in product leadership and market competitiveness. At the same time, product commercialization is essential for the growth and sustainability of any industry. As a result, industries are adapting to a fast paced approach to product development followed by an effective and streamlined commercialization process. These processes involve developments and activities connecting different business functions. This talk provides insight into the processes involved, the challenges, the partnerships, and other related activities pertaining to the progression from Idea to a product to commercialization.

# Quantitative Framework to Guide Process Development in the Pharmaceutical Industry

**Conference:** Process Development Symposium

**Year:** 2021

**Proceeding:** 2021 Process Development Symposium

**Group:** General Submissions

**Session:** Efficient Process Development Management

**Time:** Wednesday, June 9, 2021 - 12:00pm-12:30pm

## Authors:

**Rogers, A. - Presenter**, Bristol-Myers Squibb Co.

Innovation and Technology drives the commercialization process that results in product leadership and market competitiveness. At the same time, product commercialization is essential for the growth and sustainability of any industry. As a result, industries are adapting to a fast paced approach to product development followed by an effective and streamlined commercialization process. These processes involve developments and activities connecting different business functions. This talk provides insight into the processes involved, the challenges, the partnerships, and other related activities pertaining to the progression from Idea to a product to commercialization.

# Keys to Successful Contract Research from the Contractor's Perspective

**Conference:** Process Development Symposium

**Year:** 2021

**Proceeding:** 2021 Process Development Symposium

**Group:** General Submissions

**Session:** Development in the Age of Tech Transfers and Outsourcing

**Time:** Wednesday, June 9, 2021 - 1:00pm-1:30pm

## Authors:

**Nunley, R. - Presenter**, MATRIC

This presentation will discuss the keys to success and potential pitfalls of contract research from the contractor's perspective. The talk is presented by the Pilot Plant Manager for MATRIC, a research firm offering process development assistance from early concept development through scale-up. The speaker has experience with clients from a wide variety of backgrounds, company sizes, and states of technology development. The talk will lead off with thoughts on understanding your objectives to help you find the right research partner. It will explore various types of partners, such as toll converters, specialized or vendor laboratories, academic research, government funded research, and full-service private research companies followed by some general thoughts that may help determine the best fit for your research. The discussion will shift to experience at MATRIC in what makes programs successful and where pitfalls may await. This includes topics such as technical collaboration, understanding business and technical objectives, schedule, contractual expectations, funding, and scale-up.

# Effective Process Technology Transfer to Contract Manufacturers

**Conference:** Process Development Symposium

**Year:** 2021

**Proceeding:** 2021 Process Development Symposium

**Group:** General Submissions

**Session:** Development in the Age of Tech Transfers and Outsourcing

**Time:** Wednesday, June 9, 2021 - 1:30pm-2:00pm

## Authors:

**Legge, J. - Presenter**, The Dow Chemical Company

As manufacturing outsourcing becomes more prevalent in the chemical industry, The Dow Chemical Company has developed processes to effectively manage and protect intellectual property, know-how and trade secrets as part of the outsourcing process. At the same time, detailed processing information, including environmental, health, safety and quality information, must be provided by the outsourcing company to ensure successful technology transfer and safe operations at the contractor facility.

In this presentation, the general process for selecting an appropriate contract manufacturer, setting up Confidentiality Disclosure Agreements, developing Technology Transfer documents, conducting a formal Request for Quote process, supporting the contract manufacturer through scale-up/qualification and negotiating a suitable Contract Manufacturing Agreement will be discussed. For each stage, general principles leading to project success for both the company and the contractor will be provided, including best practices and pitfalls. This stepwise process should foster collaboration and trust between the company and the contractor, hopefully leading to a successful process technology transfer and the basis for a long-term contracting relationship.

# Controlling Second Phase Fall-out from a Homogeneous Mixing Process

**Conference:** Process Development Symposium

**Year:** 2021

**Proceeding:** 2021 Process Development Symposium

**Group:** General Submissions

**Session:** Development in the Age of Tech Transfers and Outsourcing

**Time:** Wednesday, June 9, 2021 - 2:00pm-2:30pm

## Authors:

**Basgall, L. - Presenter**, The Dow Chemical Company

In the production of a critical intermediate, a hydrophobic diluent and an amphiphilic active species are mixed, causing an immiscible second phase to fall out of solution. Critical to product quality, the intermediate must be a clear solution free of second phase material. Scaled-down lab work was completed to optimize the mixing and separation process and identify the amount and character of the second phase. As expected, mixing intensity and order of addition played an important role in separability of the second phase. Unexpectedly, ingress of a small amount of water was also shown to be critical to final product quality. This knowledge was key to explain past observations in production, and the potential of a reappearing second phase if the process is mishandled.

# Xolography: A linear, volumetric technology for rapid, multiscale 3D printing

**Conference:** Process Development Symposium

**Year:** 2021

**Proceeding:** 2021 Process Development Symposium

**Group:** General Submissions

**Session:** 3D Printing

**Time:** Thursday, June 10, 2021 - 8:00am-8:30am

## Authors:

**Regehly, M. - Presenter**, Xolo GmbH

Additive manufacturing is a new manufacturing technology to produce parts which are used in both, research and production plants at BASF. The focus is to use the new freedom of design to facilitate a more effective and efficient way of producing chemicals. The BASF division Global Engineering Services built an Additive Manufacturing Center and set up a team who has in the past 6 years been fabricating components with a main focus on metal alloys. Currently the material focus is on stainless steel and corrosive-resistant nickel-based alloys.

In process industry, pressure equipment poses particularly high technical requirements in terms of component integrity, safety and certification.

Therefore, certification is crucial for additive manufacturing to become an accepted manufacturing technology for pressure equipment.

The Technical Inspection at BASF SE in their role as Notified Body (user inspectorate) has certified an additively manufactured autoclave (PS: 225 bar, TS: 300°C) as a pressure equipment for the first time. This equipment conforms to the European Pressure Equipment Directive (2014/68/EU).

In this talk, the presenter will give an overview about the established Additive Manufacturing Center and the first operational experiences with pressure equipment. In addition, the range of available material and their qualification approaches will be introduced.

# Metal AM Serial Production by Cold Metal Fusion

**Conference:** Process Development Symposium

**Year:** 2021

**Proceeding:** 2021 Process Development Symposium

**Group:** General Submissions

**Session:** 3D Printing

**Time:** Thursday, June 10, 2021 - 7:30am-8:00am

## Authors:

**Staudigel, C. - Presenter**, Headmade Materials

Additive manufacturing is a new manufacturing technology to produce parts which are used in both, research and production plants at BASF. The focus is to use the new freedom of design to facilitate a more effective and efficient way of producing chemicals. The BASF division Global Engineering Services built an Additive Manufacturing Center and set up a team who has in the past 6 years been fabricating components with a main focus on metal alloys. Currently the material focus is on stainless steel and corrosive-resistant nickel-based alloys.

In process industry, pressure equipment poses particularly high technical requirements in terms of component integrity, safety and certification.

Therefore, certification is crucial for additive manufacturing to become an accepted manufacturing technology for pressure equipment.

The Technical Inspection at BASF SE in their role as Notified Body (user inspectorate) has certified an additively manufactured autoclave (PS: 225 bar, TS: 300°C) as a pressure equipment for the first time. This equipment conforms to the European Pressure Equipment Directive (2014/68/EU).

In this talk, the presenter will give an overview about the established Additive Manufacturing Center and the first operational experiences with pressure equipment. In addition, the range of available material and their qualification approaches will be introduced.

# Additive Manufacturing of Pressure Equipment

**Conference:** Process Development Symposium

**Year:** 2021

**Proceeding:** 2021 Process Development Symposium

**Group:** General Submissions

**Session:** 3D Printing

**Time:** Thursday, June 10, 2021 - 7:00am-7:30am

## Authors:

**Urbanczyk, D. - Presenter**, BASF
**Wagner, M. - Presenter**, BASF

Additive manufacturing is a new manufacturing technology to produce parts which are used in both, research and production plants at BASF. The focus is to use the new freedom of design to facilitate a more effective and efficient way of producing chemicals. The BASF division Global Engineering Services built an Additive Manufacturing Center and set up a team who has in the past 6 years been fabricating components with a main focus on metal alloys. Currently the material focus is on stainless steel and corrosive-resistant nickel-based alloys.

In process industry, pressure equipment poses particularly high technical requirements in terms of component integrity, safety and certification.

Therefore, certification is crucial for additive manufacturing to become an accepted manufacturing technology for pressure equipment.

The Technical Inspection at BASF SE in their role as Notified Body (user inspectorate) has certified an additively manufactured autoclave (PS: 225 bar, TS: 300°C) as a pressure equipment for the first time. This equipment conforms to the European Pressure Equipment Directive (2014/68/EU).

In this talk, the presenter will give an overview about the established Additive Manufacturing Center and the first operational experiences with pressure equipment. In addition, the range of available material and their qualification approaches will be introduced.

# Small changes, Big impact - How to embed safety into rapid development and scale-up of processes

**Conference:** Process Development Symposium

**Year:** 2021

**Proceeding:** 2021 Process Development Symposium

**Group:** General Submissions

**Session:** Safety in Process Development

**Time:** Thursday, June 10, 2021 - 10:00am-10:30am

## Authors:

**Roth, M. - Presenter**, Merck & Co., Inc

Additive manufacturing is a new manufacturing technology to produce parts which are used in both, research and production plants at BASF. The focus is to use the new freedom of design to facilitate a more effective and efficient way of producing chemicals. The BASF division Global Engineering Services built an Additive Manufacturing Center and set up a team who has in the past 6 years been fabricating components with a main focus on metal alloys. Currently the material focus is on stainless steel and corrosive-resistant nickel-based alloys.

In process industry, pressure equipment poses particularly high technical requirements in terms of component integrity, safety and certification.

Therefore, certification is crucial for additive manufacturing to become an accepted manufacturing technology for pressure equipment.

The Technical Inspection at BASF SE in their role as Notified Body (user inspectorate) has certified an additively manufactured autoclave (PS: 225 bar, TS: 300°C) as a pressure equipment for the first time. This equipment conforms to the European Pressure Equipment Directive (2014/68/EU).

In this talk, the presenter will give an overview about the established Additive Manufacturing Center and the first operational experiences with pressure equipment. In addition, the range of available material and their qualification approaches will be introduced.

# Plant be Nimble. Balancing Safety and Speed at the Pilot Plant Scale

**Conference:** Process Development Symposium

**Year:** 2021

**Proceeding:** 2021 Process Development Symposium

**Group:** General Submissions

**Session:** Safety in Process Development

**Time:** Thursday, June 10, 2021 - 9:30am-10:00am

## Authors:

**Van Regenmorter, J. - Presenter**, Corteva Agriscience

Additive manufacturing is a new manufacturing technology to produce parts which are used in both, research and production plants at BASF. The focus is to use the new freedom of design to facilitate a more effective and efficient way of producing chemicals. The BASF division Global Engineering Services built an Additive Manufacturing Center and set up a team who has in the past 6 years been fabricating components with a main focus on metal alloys. Currently the material focus is on stainless steel and corrosive-resistant nickel-based alloys.

In process industry, pressure equipment poses particularly high technical requirements in terms of component integrity, safety and certification.

Therefore, certification is crucial for additive manufacturing to become an accepted manufacturing technology for pressure equipment.

The Technical Inspection at BASF SE in their role as Notified Body (user inspectorate) has certified an additively manufactured autoclave (PS: 225 bar, TS: 300°C) as a pressure equipment for the first time. This equipment conforms to the European Pressure Equipment Directive (2014/68/EU).

In this talk, the presenter will give an overview about the established Additive Manufacturing Center and the first operational experiences with pressure equipment. In addition, the range of available material and their qualification approaches will be introduced.

# Enablement of Highly Hazardous Processes using Continuous Flow Technology

**Conference:** Process Development Symposium

**Year:** 2021

**Proceeding:** 2021 Process Development Symposium

**Group:** General Submissions

**Session:** Safety in Process Development

**Time:** Thursday, June 10, 2021 - 9:00am-9:30am

## Authors:

**Nagy, K. - Presenter**, DuPont Industrial Biosciences

Additive manufacturing is a new manufacturing technology to produce parts which are used in both, research and production plants at BASF. The focus is to use the new freedom of design to facilitate a more effective and efficient way of producing chemicals. The BASF division Global Engineering Services built an Additive Manufacturing Center and set up a team who has in the past 6 years been fabricating components with a main focus on metal alloys. Currently the material focus is on stainless steel and corrosive-resistant nickel-based alloys.

In process industry, pressure equipment poses particularly high technical requirements in terms of component integrity, safety and certification.

Therefore, certification is crucial for additive manufacturing to become an accepted manufacturing technology for pressure equipment.

The Technical Inspection at BASF SE in their role as Notified Body (user inspectorate) has certified an additively manufactured autoclave (PS: 225 bar, TS: 300°C) as a pressure equipment for the first time. This equipment conforms to the European Pressure Equipment Directive (2014/68/EU).

In this talk, the presenter will give an overview about the established Additive Manufacturing Center and the first operational experiences with pressure equipment. In addition, the range of available material and their qualification approaches will be introduced.

# Scale-Up of Particulate Processes - Lessons from Industry

**Conference:** Process Development Symposium

**Year:** 2021

**Proceeding:** 2021 Process Development Symposium

**Group:** General Submissions

**Session:** Scale-Up & Scale-Down/Risk Management for Accelerated Process Development

**Time:** Thursday, June 10, 2021 - 11:00am-11:30am

## Authors:

**Bell, T. A. - Presenter**, DuPont (retired)

Additive manufacturing is a new manufacturing technology to produce parts which are used in both, research and production plants at BASF. The focus is to use the new freedom of design to facilitate a more effective and efficient way of producing chemicals. The BASF division Global Engineering Services built an Additive Manufacturing Center and set up a team who has in the past 6 years been fabricating components with a main focus on metal alloys. Currently the material focus is on stainless steel and corrosive-resistant nickel-based alloys.

In process industry, pressure equipment poses particularly high technical requirements in terms of component integrity, safety and certification.

Therefore, certification is crucial for additive manufacturing to become an accepted manufacturing technology for pressure equipment.

The Technical Inspection at BASF SE in their role as Notified Body (user inspectorate) has certified an additively manufactured autoclave (PS: 225 bar, TS: 300°C) as a pressure equipment for the first time. This equipment conforms to the European Pressure Equipment Directive (2014/68/EU).

In this talk, the presenter will give an overview about the established Additive Manufacturing Center and the first operational experiences with pressure equipment. In addition, the range of available material and their qualification approaches will be introduced.

# Immiscible Liquid-Liquid droplet dynamics – A CFD model for robust scale-up and risk mitigation

**Conference:** Process Development Symposium

**Year:** 2021

**Proceeding:** 2021 Process Development Symposium

**Group:** General Submissions

**Session:** Scale-Up & Scale-Down/Risk Management for Accelerated Process Development

**Time:** Thursday, June 10, 2021 - 11:30am-12:00pm

## Authors:

**Wutz, J. - Presenter**, Boehringer-Ingelheim

Liquid-liquid systems have a wide range of applications in many fields of the industry. In chemical manufacturing, such liquid immiscibility is most commonly utilized for the purpose of chemical separations by liquid-liquid extraction. Such interphase phenomena depend highly on the droplet size distribution.

The knowledge gained by CFD can be used as a basis for a physics-driven scale-up. The recent years have seen a rapid development of new Graphics Processing Units(GPUs) for computational sciences. We leverage this technical progress with the application of Lattice-Boltzmann (LB) methods which means that we do not longer have to rely on Reynolds Averaged Navier-Stakes equations (RANS).

We begin by performing high resolution direct simulation of two immiscible fluids blending in a lab-scale agitated vessel(Figure 1).

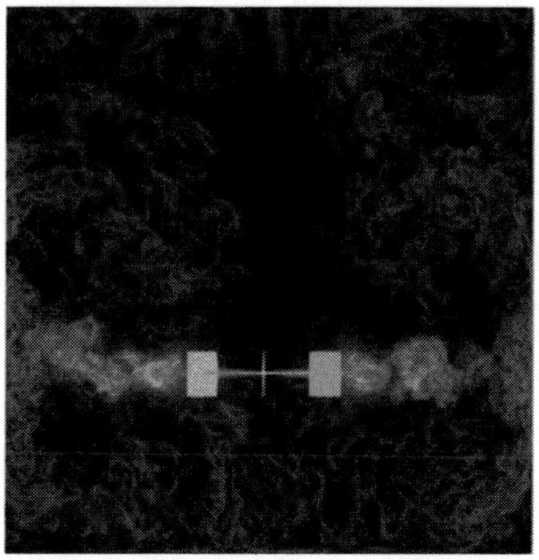

We compare droplet size predictions from this model to first-principles turbulence theory, as well as experimentally measured droplet data. We then use this direct model to simulate the break-up of

individual droplets moving through the impeller trailing vortex. From these predictions, we develop a predictive relationship between energy dissipation rates and break-up volume fraction.

We can extract the breakup characteristics of these two-phase immiscible direct simulations and use this knowledge for the design of a model that can be used for larger volumes. The direct simulation provides data on the size and position of the droplets and by detecting step changes in the droplet volume, we can isolate break-up events (Figure 2).

Then, we use a Large Eddy turbulence model (LES) in combination with Lagrangian particles as the dispersed phase to reproduce the droplet size distribution from the direct method with this much faster setup. We are using a hard core-soft shell approach for the particles in this Eulerian/Lagrangian/DEM hybrid model (Figure 3).

This approach allows us to accurately predict droplet size distributions in larger scales in an efficient and fast CFD environment. In order to address full scale production vessels, we use a parcel approach based on the Eulerian/Lagrangian/DEM hybrid model.

All three models predict similar droplet size distribution while each abstraction speeds up the simulation time significantly.

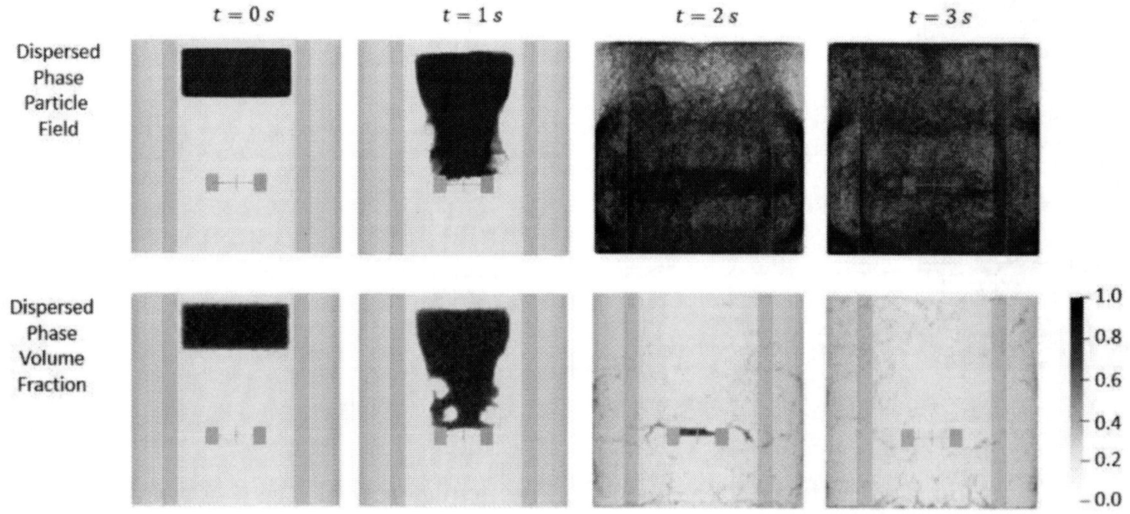

# Features of ProSim's BatchReactor software for Scale-up & Scale-down

**Conference:** Process Development Symposium

**Year:** 2021

**Proceeding:** 2021 Process Development Symposium

**Group:** General Submissions

**Session:** Scale-Up & Scale-Down/Risk Management for Accelerated Process Development

**Time:** Thursday, June 10, 2021 - 12:00pm-12:15pm

## Authors:

**Wincure, B. - Presenter**, ProSim, Inc.

ProSim's BatchReactor® software [1] is dedicated to the simulation of batch chemical reactors, which provides the user with a detailed modeling of the reactor (heating/cooling system, condenser, mixing device... ), a reliable description of the production recipe, and an analysis of scale-up or scale-down based on the volume or diameter of the reactor, for a reactor with uniform mixing (perfectly mixed). BatchReactor® allows modeling of complex systems by taking advantage of the power of Simulis® Thermodynamics (thermodynamic properties server) which offers an extensive set of thermodynamic models and a pure components database of more than 2,300 pure components (AIChE's DIPPR® database [2]). Moreover, thanks to the user "interpreted" kinetic rate model available through Simulis® Reaction (chemical reaction server), any kind of specific reaction models can be used in BatchReactor®, including classical models (equilibrium, kinetic, instantaneous) and bio models (Monod, Hill, Comtois, Competitive inhibition, Uncompetitive inhibition, Non-competitive Inhibition, etc). This is essential to represent kinetic bio-reactions with often non-classical laws [3]. So, in BatchReactor®, a stoichio-kinetic model can be coupled with thermodynamic and heat transfer models already available in the software to simulate the bio-reactor.

The strengths of the BatchReactor simulation are accurate thermodynamic and transfer phenomena representation, classical and bio reactions, recipe description, and a user-friendly interface.

Keywords

Batch reaction; Bio-reaction; Modeling; Simulation; Thermodynamic server; Chemical reaction server

References

[1] www.prosim.net

[2] R.L. Rowley, W.V. Wilding, J.L. Oscarson, N.F. Giles, DIPPR® Data Compilation of Pure Chemical Properties, Design Institute for Physical Properties, AIChE, New York, NY (2011)

[3] Ramon Portugal F., Fillon M., Meyer X.M., Pingaud H., Strehiano P., I.S.B.N. 2-87805-022-3, 159-170 (1997)

# Accelerated Development of Processes: Case study of silicon-to-operations scale-up and risk management

**Conference:** Process Development Symposium

**Year:** 2021

**Proceeding:** 2021 Process Development Symposium

**Group:** General Submissions

**Session:** Scale-Up & Scale-Down/Risk Management for Accelerated Process Development

**Time:** Thursday, June 10, 2021 - 12:15pm-12:45pm

## Authors:

**Skoulidas, A. - Presenter**, ExxonMobil

ProSim's BatchReactor® software [1] is dedicated to the simulation of batch chemical reactors, which provides the user with a detailed modeling of the reactor (heating/cooling system, condenser, mixing device... ), a reliable description of the production recipe, and an analysis of scale-up or scale-down based on the volume or diameter of the reactor, for a reactor with uniform mixing (perfectly mixed). BatchReactor® allows modeling of complex systems by taking advantage of the power of Simulis® Thermodynamics (thermodynamic properties server) which offers an extensive set of thermodynamic models and a pure components database of more than 2,300 pure components (AIChE's DIPPR® database [2]). Moreover, thanks to the user "interpreted" kinetic rate model available through Simulis® Reaction (chemical reaction server), any kind of specific reaction models can be used in BatchReactor®, including classical models (equilibrium, kinetic, instantaneous) and bio models (Monod, Hill, Comtois, Competitive inhibition, Uncompetitive inhibition, Non-competitive Inhibition, etc). This is essential to represent kinetic bio-reactions with often non-classical laws [3]. So, in BatchReactor®, a stoichio-kinetic model can be coupled with thermodynamic and heat transfer models already available in the software to simulate the bio-reactor.

The strengths of the BatchReactor simulation are accurate thermodynamic and transfer phenomena representation, classical and bio reactions, recipe description, and a user-friendly interface.

Keywords

Batch reaction; Bio-reaction; Modeling; Simulation; Thermodynamic server; Chemical reaction server

References

[1] www.prosim.net

[2] R.L. Rowley, W.V. Wilding, J.L. Oscarson, N.F. Giles, DIPPR® Data Compilation of Pure Chemical Properties, Design Institute for Physical Properties, AIChE, New York, NY (2011)

[3] Ramon Portugal F., Fillon M., Meyer X.M., Pingaud H., Strehiano P., I.S.B.N. 2-87805-022-3, 159-170 (1997)

# Scaling-up From High Throughput Screening to Continuous Process for the Synthesis of 1,6-Hexanediol from Renewable Feedstocks

**Conference:** Process Development Symposium

**Year:** 2021

**Proceeding:** 2021 Process Development Symposium

**Group:** General Submissions

**Session:** Batch vs. Continuous

**Time:** Thursday, June 10, 2021 - 1:30pm-2:00pm

## Authors:

**Sengupta, S. - Presenter**, DuPont

Since the beginning of the 21st century, the chemical industry has been making many novel and ingenious innovations to design cost-advantaged, inherently safe, and sustainable processes based on renewably- sourced feedstocks for the production of fuels, chemicals, and advanced materials. However, there is still a persistent public perception that the chemical industry is not doing enough to develop safe and sustainable chemistry. As a matter of fact, a large number of scientists, environmental scholars, business leaders, and governmental policymakers believe that chemical industry must do more to develop sustainable processes to save our planet from the dire consequences of climate change.

During the last three decades, DuPont has been in the forefront of developing sustainable products and processes. One such endeavor was to develop a cost-competitive process for the synthesis of 1,6-hexanediol from renewable feedstocks. Several potential feedstocks have been investigated with the main intermediates being 1,2,6-hexanetriol (1,2,6-HT) and tetrahydropyran-2-methanol (THP2M) as shown in Figure 1.

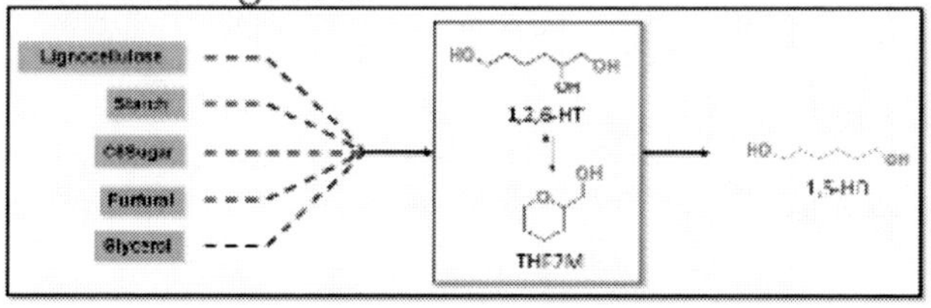


Figure 1: Reactions highlighting the main intermediates for converting biomass-derived feedstocks to renewable 1,2-hexanediol

Multimetallic catalysts have been shown to be valuable for a variety of industrially-practiced reactions, including biomass valorization. We have explored the activity, selectivity, and life of supported metal-metal oxide catalyst combinations using high throughput screening and continuous trickle bed reactors for the hydrodeoxygenation of biomass-derived feedstocks. [1-5]

The development work started with the synthesis and screening of plethora of supported metal and metal-oxide catalyst systems (Figure 2). The initial proof of concept (PoC) was performed in small high- throughput batch reactors (milligram quantity of catalysts in powder form). After completing

PoC, the process was transferred from batch reactors to continuous reactors, which comprised of high-throughput trickle-bed test units (gram catalyst scale, in powder form, with 16 reactor tubes in parallel) to closely match the operating conditions of a commercial continuous hydrodeoxygenation process. The first two steps comprised of screening several hundred catalysts under a broad range of operating conditions to determine the activity, selectivity, and stability of the catalysts. Performance targets were set based on the initial results and techno-economic evaluation. In the next step, several lead catalyst candidates were identified and subsequently scaled-up to shapes (extrudates and pellets) and sizes (e.g., 1/16â€ and 1/8â€ )
similar to that of standard commercial catalysts. During this evaluation process, a few grams of the selected catalysts were further optimized by extending the range of operating conditions. In the final stage of the development work, the amount of catalyst was increased into hundred-gram scale using a semiworks-scale test rig. The latter had a lower degree of parallelization but a high degree of flexibility for each catalyst system. The catalyst systems were operated independently in their preferred reaction conditions and over a longer period with industrially relevant feedstock to assess long-term catalyst stability. Larger quantities of the effluent streams were collected for more detailed product analytics and separation and purification workstreams. The results obtained from the reactor scale-up study were used to further refine the techno-economic analysis and were used as an input for designing the pilot plant.

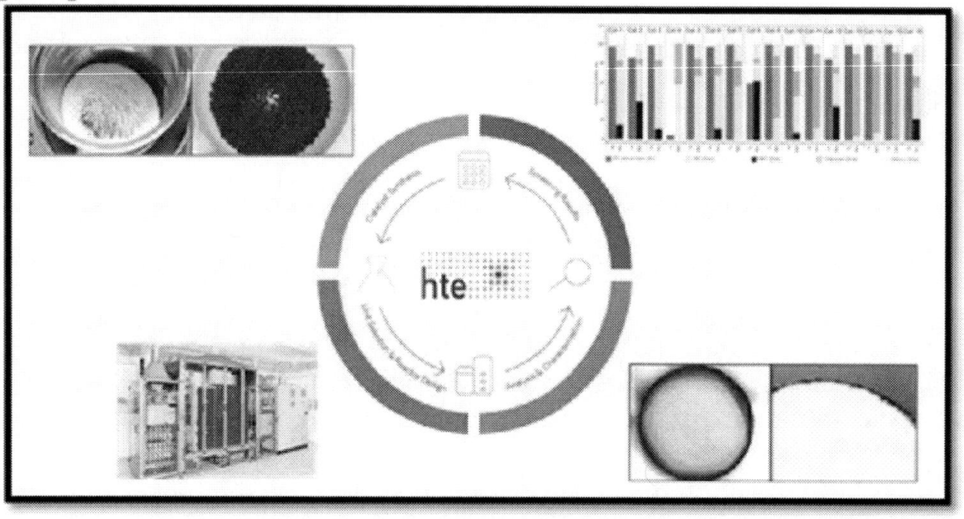


Figure 2: Development cycle for producing 1,6-HD from C6 polyols. The catalyst scale-up in this example involved shell impregnation of commercial supports. The product analysis involved detailed GC-MS and GC-FID/TCD analyses of the gas- and liquid-phase products. The testing included batch and continuous high throughput screening and flexible semiworks-scale reactor system.

The performance of the catalysts, with special reference to Pt-WOx/TiO2, the lead catalyst candidate, for the hydrodeoxygenation of 1,2-HT and THP2M and the significance of the batch and continuous reactor systems for the synthesis of 1,6-HD will be discussed in the presentation.

# References

1. Allgeier, A. M., Carlson; T. R., Corbin, D. R., De Silva, W. I. N., Menning, C. A., Ritter, J. C., and Sengupta, S. K., Production of alpha, omega-diols US 9,670,118 (2017)
2. Allgeier, A. M., Carlson; T. R., Corbin, D. R., De Silva, W. I. N., Korovessi, E., Menning, C. A., Ritter, J. C., Rosenfeld; H. D. and Sengupta, S. K., Production of alpha, omega-diols US 9,181,157 (2015)
3. Stauffer, C. S., Sengupta, S. K., Ritter, J. C., Menning, C. A., Korovessi, E., De Silva, W. I. N., Allgeier,
A. M., Process for Preparing 1,6-Hexanediol, US 8,865,940 (2014)

4. Stauffer, C. S., Sengupta, S. K., Ritter, J. C., Menning, C. A., Korovessi, E., De Silva, W. I. N., Allgeier, A. M., Process for Preparing 1,6-Hexanediol, US 8,865,940 (2014)
5. Stephens, K. J., Allgeier, A. M., Bell, A. L., Carlson, T. R., Chen, Y., Douglas, J. T., Howe, L. A., Menning, C. A., Neuenswander, S. A., Sengupta, S. K., Thapa, and Ritter, J. C. P. S. A Mechanistic Study of Polyol Hydrodeoxygenation over a Bifunctional Pt-WOX/TiO2 Catalyst, ACS Catal. 10, 12996â^'13007 (2020)

# Going with the Flow: Performance Drivers for Pharmaceutical Development and How Flow Chemistry Fits into the Mix

**Conference:** Process Development Symposium

**Year:** 2021

**Proceeding:** 2021 Process Development Symposium

**Group:** General Submissions

**Session:** Batch vs. Continuous

**Time:** Thursday, June 10, 2021 - 1:00pm-1:30pm

## Authors:

**Sacia, E. R. - Presenter**, University of California, Berkeley

Flow processes offer differentiated advantages in their ability to carry out challenging chemical reactions safer and often faster, with smaller footprint, and with more consistent quality. However, much of pharmaceutical development is still performed in batch processing due to constraints of existing infrastructure, convention, perceived regulatory restrictions, and availability of adequate Engineering resources to support the scale-up challenges. In this talk, two case studies will be provided to discuss advantages and opportunities for pharmaceutical processes executed in flow vs. batch.

In the first case study, the emerging use of visible light as a "reagent" for photochemically-driven organic reactions will be discussed. Photochemical methodologies enable complex organic molecules to be synthesized with catalysts that can facilitate single-electron transfer from the catalyst's excited state. Such reactions are often difficult to selectively enable through more conventional two-electron processes. However, to develop and scale-up such syntheses, appropriate chemical reactors must be designed and validated to enable demonstration and commercial production at manufacturing sites. For this purpose, flow reactors can provide an attractive advantage to improve reactor productivity, photon flux, heat transport, and mass transport. For this reason, a laser-driven continuously-stirred tank reactor (CSTR) has been successfully implemented for developing and scaling photocatalytic chemistries.

In the second case study, the role of reactor design on selective hydrogenation chemistry will be discussed. Reactor design and operating parameters (e.g. – pressure, temperature, gas-to-liquid ratio, catalyst loading, and residence time) may have a profound impact on substrate enantioselectivity and reactor robustness. While hydrogenations in flow are ubiquitous in the petroleum industry, their adoption in the pharmaceutical industry has not been nearly as widespread. For this reason, we will describe how fit-for-purpose modeling may be used to rapidly identify optimal conditions and configurations for flow hydrogenation development and why they can be advantaged over batch operation.

Finally, a brief perspective will be provided on barriers and scale-up challenges for processes in the pharmaceutical industry. This review will cover the constraints and areas of growth for future processing capabilities in batch and flow within the sector.

# Batch vs. Continuous Operation in Solid-Liquid Separations

**Conference:** Process Development Symposium

**Year:** 2021

**Proceeding:** 2021 Process Development Symposium

**Group:** General Submissions

**Session:** Batch vs. Continuous

**Time:** Thursday, June 10, 2021 - 2:00pm-2:30pm

## Authors:

**Wolff, S. H. - Presenter**, DuPont

Since the beginning of the 21st century, the chemical industry has been making many novel and ingenious innovations to design cost-advantaged, inherently safe, and sustainable processes based on renewably- sourced feedstocks for the production of fuels, chemicals, and advanced materials. However, there is still a persistent public perception that the chemical industry is not doing enough to develop safe and sustainable chemistry. As a matter of fact, a large number of scientists, environmental scholars, business leaders, and governmental policymakers believe that chemical industry must do more to develop sustainable processes to save our planet from the dire consequences of climate change.
During the last three decades, DuPont has been in the forefront of developing sustainable products and processes. One such endeavor was to develop a cost-competitive process for the synthesis of 1,6-hexanediol from renewable feedstocks. Several potential feedstocks have been investigated with the main intermediates being 1,2,6-hexanetriol (1,2,6-HT) and tetrahydropyran-2-methanol (THP2M) as shown in Figure 1.

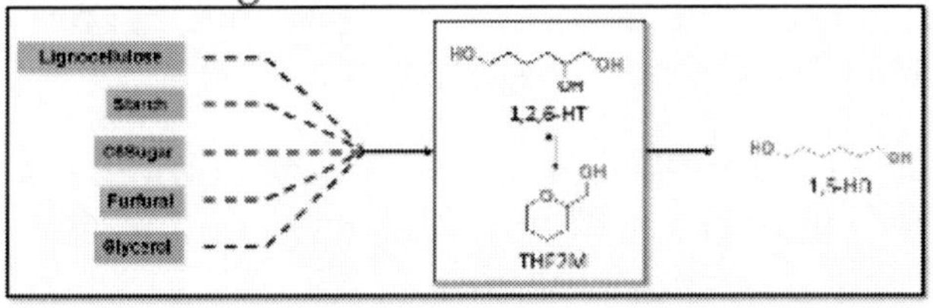


Figure 1: Reactions highlighting the main intermediates for converting biomass-derived feedstocks to renewable 1,2-hexanediol

Multimetallic catalysts have been shown to be valuable for a variety of industrially-practiced reactions, including biomass valorization. We have explored the activity, selectivity, and life of supported metal-metal oxide catalyst combinations using high throughput screening and continuous trickle bed reactors for the hydrodeoxygenation of biomass-derived feedstocks. [1-5]
The development work started with the synthesis and screening of plethora of supported metal and metal-oxide catalyst systems (Figure 2). The initial proof of concept (PoC) was performed in small high- throughput batch reactors (milligram quantity of catalysts in powder form). After completing PoC, the process was transferred from batch reactors to continuous reactors, which comprised of high-throughput trickle-bed test units (gram catalyst scale, in powder form, with 16 reactor tubes in

parallel) to closely match the operating conditions of a commercial continuous hydrodeoxygenation process. The first two steps comprised of screening several hundred catalysts under a broad range of operating conditions to determine the activity, selectivity, and stability of the catalysts. Performance targets were set based on the initial results and techno-economic evaluation. In the next step, several lead catalyst candidates were identified and subsequently scaled-up to shapes (extrudates and pellets) and sizes (e.g., 1/16â€ and 1/8â€ )
similar to that of standard commercial catalysts. During this evaluation process, a few grams of the selected catalysts were further optimized by extending the range of operating conditions. In the final stage of the development work, the amount of catalyst was increased into hundred-gram scale using a semiworks-scale test rig. The latter had a lower degree of parallelization but a high degree of flexibility for each catalyst system. The catalyst systems were operated independently in their preferred reaction conditions and over a longer period with industrially relevant feedstock to assess long-term catalyst stability. Larger quantities of the effluent streams were collected for more detailed product analytics and separation and purification workstreams. The results obtained from the reactor scale-up study were used to further refine the techno-economic analysis and were used as an input for designing the pilot plant.

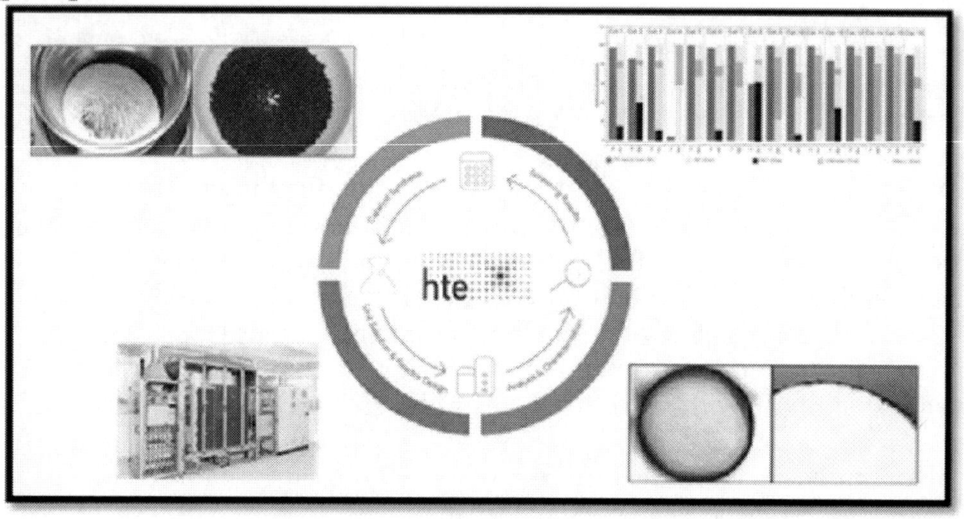


Figure 2: Development cycle for producing 1,6-HD from C6 polyols. The catalyst scale-up in this example involved shell impregnation of commercial supports. The product analysis involved detailed GC-MS and GC-FID/TCD analyses of the gas- and liquid-phase products. The testing included batch and continuous high throughput screening and flexible semiworks-scale reactor system.

The performance of the catalysts, with special reference to Pt-WOx/TiO2, the lead catalyst candidate, for the hydrodeoxygenation of 1,2-HT and THP2M and the significance of the batch and continuous reactor systems for the synthesis of 1,6-HD will be discussed in the presentation.

# References

1. Allgeier, A. M., Carlson; T. R., Corbin, D. R., De Silva, W. I. N., Menning, C. A., Ritter, J. C., and Sengupta, S. K., Production of alpha, omega-diols US 9,670,118 (2017)
2. Allgeier, A. M., Carlson; T. R., Corbin, D. R., De Silva, W. I. N., Korovessi, E., Menning, C. A., Ritter, J. C., Rosenfeld; H. D. and Sengupta, S. K., Production of alpha, omega-diols US 9,181,157 (2015)
3. Stauffer, C. S., Sengupta, S. K., Ritter, J. C., Menning, C. A., Korovessi, E., De Silva, W. I. N., Allgeier,
A. M., Process for Preparing 1,6-Hexanediol, US 8,865,940 (2014)
4. Stauffer, C. S., Sengupta, S. K., Ritter, J. C., Menning, C. A., Korovessi, E., De Silva, W. I. N., Allgeier, A. M., Process for Preparing 1,6-Hexanediol, US 8,865,940 (2014)

5. Stephens, K. J., Allgeier, A. M., Bell, A. L., Carlson, T. R., Chen, Y., Douglas, J. T., Howe, L. A., Menning, C. A., Neuenswander, S. A., Sengupta, S. K., Thapa, and Ritter, J. C. P. S. A Mechanistic Study of Polyol Hydrodeoxygenation over a Bifunctional Pt-WOX/TiO2 Catalyst, ACS Catal. 10, 12996â^'13007 (2020)

# AUTHOR INDEX